EXTREME ENVIRONMENTAL THREATS™

CLIMATE CHANGE

Human Effects on the Nitrogen Cycle

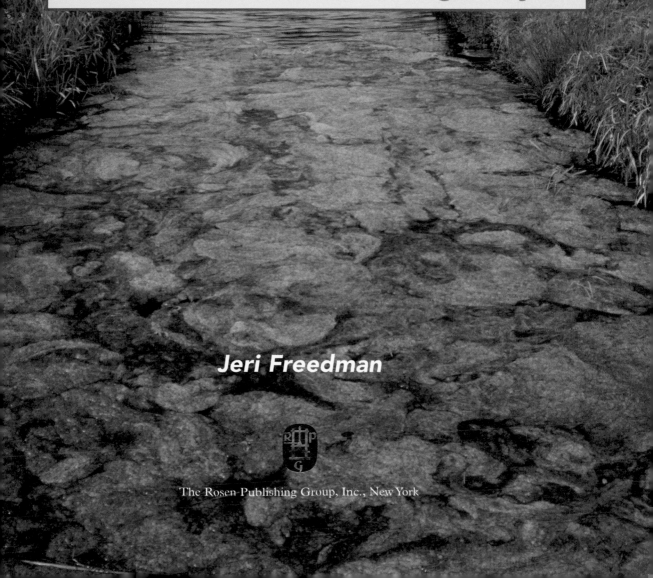

Jeri Freedman

The Rosen Publishing Group, Inc., New York

To my niece and nephew, Laura and Matthew Freedman, with love

Published in 2007 by The Rosen Publishing Group, Inc.
29 East 21st Street, New York, NY 10010

First Edition

Library of Congress Cataloging-in-Publication Data

Freedman, Jeri.
Climate change: human effects on the nitrogen cycle/Jeri Freedman.—1st ed.
    p. cm.—(Extreme environmental threats)
Includes bibliographical references and index.
ISBN 1-4042-0744-9 (lib. bdg.)
1. Nitrogen cycle. 2. Global warming. 3. Nature—Effect of human beings on.
I. Title. II. Series.
QH344.F74 2007
577.27'6—dc22

                                                              2006000164

Manufactured in the United States of America

On the cover: An aquatic plant floats on the surface of Lake Maracaibo in the state of Zulia, Venezuela, approximately 300 miles (500 kilometers) from Caracas. In June 2004, the growth covered 10 to 20 percent of the lake's 5,400-square-mile (13,500-square-kilometer) surface. An excess of fertilizer runoff, washed into the lake by heavy rains, was the likely cause of the extreme growth. **Title page:** This canal, which runs through farmland in Germany, is filled with algae as a result of runoff from nitrogen-containing fertilizer that has been used in the surrounding fields.

Contents

INTRODUCTION

Ammonium nitrate fertilizer, seen here, is being poured into a railway car for shipment.

Nitrogen, not oxygen, is the most common gas in our atmosphere. Nitrogen therefore plays a very important role in maintaining life on Earth. The nitrogen cycle is a series of chemical reactions through which nitrogen in the atmosphere is changed into a form that living creatures can use. It is then broken down and returned to the atmosphere.

The nitrogen cycle is one of the processes by which chemicals needed for life are continually used

and recycled. In this way, chemicals that exist in finite amounts on Earth are continuously available to living organisms.

Human population growth and activities are having a major impact on the nitrogen cycle. Since the start of the Industrial Revolution about 200 years ago, the human population has significantly increased the amount of biologically available nitrogen. This increase in nitrogen-based compounds leads to global warming and related changes in climate, sea level, and the ozone layer. It also has the potential to have great effects on both land-based and coastal ecosystems (groups of plants and animals living in a particular environment).

If global warming continues unchecked, sudden temperature changes could occur throughout the world. Temperatures could drop 5 to 6 degrees Fahrenheit (3 to 3.5 degrees Celsius) across Asia, Europe, and North America, while rising 4°F (2.5°C) in Africa, South America, and Australia.

If drought affects the biggest food-growing areas of the world, food production would decline and worldwide famine could occur for as long as a decade. The largest urban areas of the eastern United States and northern Europe could suffer severe water shortages, as could other parts of the world.

Blizzards and hurricanes could become stronger, and high winds could hammer coastal regions. Higher seas could cause an increase in the height of giant waves called tsunamis, which result from underwater earthquakes. They could rise as high as eighty feet (twenty-five meters) in the air. These waves could wash away entire island populations and destroy seaside communities around the world.

Shortages of food, water, and electricity might lead to local skirmishes and global war if desperate people who are determined to survive compete for limited resources. But could such destruction really happen in our lifetime? Some scientists believe that if we do not address the problems that produce global warming, it just might.

There are some things that people can do about rising nitrogen amounts, however. The following chapters look in detail at some of the effects on the nitrogen cycle, present scientific research, and discuss ways that the problem might be addressed in the future.

THE END OF LIFE
AS WE KNOW IT?

Workers arrive at a
chemical fertilizer
plant near
Alexandria, Egypt.

Nitrogen makes up 78 percent of the air we breathe, and it plays a critical role in life on Earth. Nitrogen is an important component of deoxyribonucleic acid (DNA), the genetic-information-carrying material in our cells. It is also part of other compounds, called proteins, used throughout our bodies.

The nitrogen in our atmosphere exists primarily as two atoms (the most basic particles that make up objects) bound together to form one nitrogen molecule (a combination of atoms). This form of nitrogen is called dinitrogen (N_2). About 78 percent of the nitrogen in our atmosphere is dinitrogen. Plants and animals are not able to use nitrogen in this form, however. Special bacteria in the soil alter the dinitrogen so that it forms compounds with other elements, such as oxygen. Plants get their nitrogen from these compounds, which are deposited in the soil. Animals, including people, get their nitrogen from eating plants and animals that have eaten plants.

Some nitrogen forms various combinations with oxygen, making nitrous oxide (N_2O), nitrite (NO_2), nitrate (NO_3), and nitric oxide (NO). Nitrogen can also be found in compounds such as ammonia (NH_3). Nitrogen is present in air, water, and solid forms. Although nitrogen sustains life on Earth, some forms of nitrogen, such as nitrous oxide, can pose a threat to life because they contribute to global warming.

For most of Earth's history, nitrogen, in a form usable for living things, was in short supply. This was because only lightning and special types of bacteria could convert it into a form that plants and animals could use. However, the development of plants and bacteria that could convert atmospheric nitrogen to usable nitrogen was an important historical factor in the increase in available nitrogen. In addition, over the past several centuries, humans have used fertilizers that contain nitrogen and burned fossil fuels such as coal and oil, which have created possibly too much nitrogen.

Nitrogen Compounds

When talking about chemical compounds, a standard form is used to show what each compound contains. All compounds are made up of combinations of basic elements. The smallest unit of any element is called an atom. When two or more atoms combine, they form a molecule, which is a single unit of a given compound. When different elements combine to form compounds, exactly what a compound consists of can be indicated by showing the number of atoms of each element that a molecule of the compound contains. For example, the compound nitrous oxide (N_2O) contains two atoms of nitrogen and one atom of oxygen. Elements you will see in this book are:

N = Nitrogen
O = Oxygen
C = Carbon
H = Hydrogen

Atoms of these elements combine to form the following compounds found in this book:

CO_2 = Carbon Dioxide
N_2 = Dinitrogen
NH_3 = Ammonia
NO = Nitric Oxide
NO_2 = Nitrite
NO_3 = Nitrate
N_2O = Nitrous Oxide
O_3 = Ozone

GREENHOUSE GASES AND THE GREEN-HOUSE EFFECT

While solar radiation gently warms the earth, too much of certain gases in the atmosphere could cause the planet to over-heat. This would lead to large changes in the cli-mate that would disrupt all life on Earth.

Most atmospheric gases do not increase global warming. These gases include dinitrogen and oxygen, which together make up 99 percent of the gases in the atmosphere.

Models of six nitrogen oxides are shown here. The blue spheres represent nitrogen atoms; the red spheres represent oxygen atoms. Most nitrogen oxides are atmos-pheric pollutants.

The remaining 1 percent of the atmosphere's gases, however, increases the greenhouse effect and therefore global warming. These gases include carbon dioxide (CO_2), ozone (O_3), and nitrous oxide (N_2O). Looking at these three greenhouses gases, you can see that they all contain three atoms. This is a very important difference from the atmosphere's simple

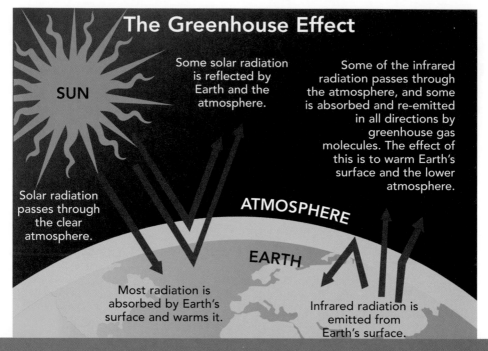

The Greenhouse Effect

SUN

Some solar radiation is reflected by Earth and the atmosphere.

Some of the infrared radiation passes through the atmosphere, and some is absorbed and re-emitted in all directions by greenhouse gas molecules. The effect of this is to warm Earth's surface and the lower atmosphere.

Solar radiation passes through the clear atmosphere.

ATMOSPHERE

EARTH

Most radiation is absorbed by Earth's surface and warms it.

Infrared radiation is emitted from Earth's surface.

This diagram shows how the greenhouse effect works. Radiation from the sun passes through the atmosphere and hits Earth's surface. Some of the radiation passes through the atmosphere and goes into space, and some of the radiation is absorbed by molecules in the atmosphere and reflected back to Earth.

oxygen (O_2) and nitrogen (N_2), which each consist of only two atoms.

GASES IN THE GREENHOUSE

If you have ever seen a rainbow, you know that sunlight is actually composed of different colors of light. The variety of colors occurs because waves of light vibrate, or radiate, at different frequencies (the number of times

Fourier in the Greenhouse

Jean-Baptiste-Joseph Fourier (1768–1830), a French mathematician, was the first scientist to recognize that the earth's atmosphere is important in keeping the planet a place where life could survive. In 1824, Fourier used the idea of a greenhouse to explain the atmosphere's effect on Earth. He said that the gases surrounding Earth were similar to the walls of a greenhouse. The gases keep in the sun's heat so that the surface of the planet is warmed to a temperature suitable for maintaining life.

In other words, Fourier was the first person to describe the greenhouse effect.

FOURIER

Baron Jean-Baptiste-Joseph Fourier was born in Auxerre, France. He studied mathematics and became a teacher. In 1798, he served as science advisor to Napoleon (1769–1821) during the invasion of Egypt. After the war, he returned to France and continued his mathematical research. He wrote a variety of papers on mathematical and scientific topics, including "On the Propagation of Heat in Solid Bodies." In 1811, he won the mathematics prize from the Paris Institute for his work on that subject.

Visible light, passing through water molecules in the atmosphere after a rainfall, can form a rainbow. The different wavelengths of visible light appear as different colors. The shortest wavelength appears as violet, while the longest is red. Beyond the visible range of light are still shorter (ultraviolet) and longer (infrared) light wavelengths. Even though we can't see them, they also interact with molecules in the atmosphere and affect life on Earth.

per second that a particular shape of wave occurs in a vibrating wave of energy).

The various colors of light that we can see with our naked eyes are called visible light. These colors are red, orange, yellow, green, blue, and violet. Red has the longest wavelength, orange the next longest, and so on, with violet having the shortest wavelength.

However, there are other wavelengths of light that we can't see without special devices. One type is infrared radiation, which is light that has a longer wavelength than the red we can see, and thus it is invisible to our eyes.

When sunlight reflects off the surface of Earth, one of two things can happen. Either it passes back through the atmosphere to outer space or it is reflected back to the surface of Earth, thus heating it up.

Two- and three-atom molecules react differently to light. Two-atom molecules allow visible and infrared light to pass through them, but three-atom molecules, such as nitrous oxide (N_2O), let only visible light pass through. Therefore, infrared radiation hits these molecules, causing their atoms to vibrate and, in the process, give off energy. This energy hits Earth in the form of more infrared radiation. When this infrared radiation bounces back from Earth to strike the three-atom molecules again, the cycle is repeated, over and over, until the radiation is scattered into space.

Some of this infrared-reflection process is good for Earth. It keeps the planet warm enough to sustain life. However, when too much of these infrared-reflecting gases accumulate in the atmosphere, so much heat is trapped that it can have serious effects on Earth's climate.

The Danube River, shown here, has been polluted by a fertilizer plant that is located along its banks in Pančevo, Yugoslavia.

One of the factors that contribute to global warming is an increase in atmospheric nitrogen (nitrogen in the air), especially nitrous oxide (N_2O). Nitrogen exists in several different forms. However, most nitrogen exists in forms that can't be used by living creatures. Nitrogen must be transformed, or changed, into a form that plants and animals can use in a process called nitrogen fixation.

In order for nitrogen to be useful to people and other living organisms, the molecules of nitrogen must be broken apart. This is done by certain bacteria, such as rhizobia, that live in soil and capture nitrogen from the air. Plants then absorb nitrogen from the soil. Human beings and other animals get their nitrogen from eating plants. When plants decay and when animals produce manure, bacteria break down these materials and return the nitrogen to the air, and the cycle starts again.

These students are examining nitrogen-producing nodules on a soybean plant. Proteins and enzymes in the plant allow it to absorb nitrogen from the atmosphere and convert it into ammonia.

However, human beings are having a major impact on the nitrogen cycle. Today, more than half of the nitrogen that can be absorbed by plants comes from nitrogen-enhanced fertilizer; about a third comes from food crops such as legumes (beans), which are one of a few types of plants that can fix nitrogen directly without it first being broken down by bacteria; and most of the rest comes from the burning of fossil fuels such as oil and coal.

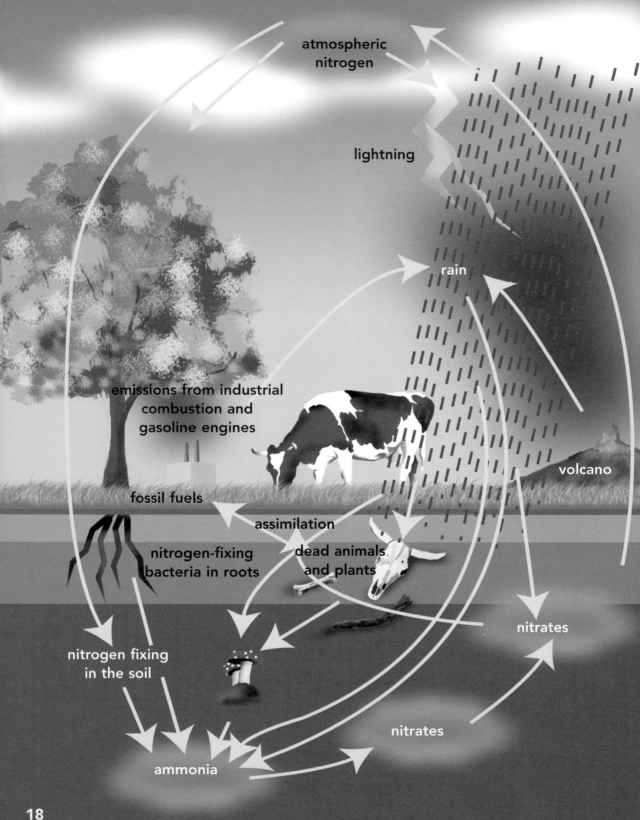

atmospheric
nitrogen

lightning

rain

emissions from industrial
combustion and
gasoline engines

volcano

fossil fuels

assimilation

nitrogen-fixing
bacteria in roots

dead animals
and plants

nitrates

nitrogen fixing
in the soil

nitrates

ammonia

ARE WE AFFECTING THE CLIMATE?

The human population is having a big effect on the planet and the weather. In December 2003, the American Geophysical Union stated, "Human activities are increasingly altering the Earth's climate . . . scientific evidence strongly indicates that natural influences cannot explain the rapid increase in global near-surface temperatures observed during the second half of the 20th century." It goes on to say, "Human impacts on the climate system include increasing concentrations of atmospheric greenhouse gases (e.g., carbon dioxide, chlorofluorocarbons and their substitutes, methane, nitrous oxide, etc.), air pollution, increasing concentrations of airborne particles, and land alteration."

In 2000, the United Nations Intergovernmental Panel on Climate Change issued a report in which it stated, "There has been a discernable human influence on global climate."

The diagram on the facing page illustrates how the nitrogen cycle works. Nitrogen (N_2) in the air passes to the soil, where bacteria transform it to ammonia (NH_3) and nitrates (NO_3), forms that can be used by plants. Animals then eat the plants, thus consuming the nitrogen. When an animal dies and decomposes, nitrogen is released and returns to the soil or air.

Melting Arctic Ice

Scientists' records over the past fifty years show that the amount of ice in the Arctic has been shrinking. It has been melting at the rate of 14,000 square miles (36,200 square kilometers) per year. Scientists have been studying this phenomenon to see if it is part of a natural cycle or the result of human activities.

In 1999, scientists at the U.S. National Aeronautics and Space Administration's (NASA) Goddard Space Flight Center in Greenbelt, Maryland, used computer models to see if natural or human activities caused the melting. They found that there was only a 2 percent chance that the melting during the last two decades of the twentieth century was solely the result of natural forces. Furthermore, there was less than a 0.1 percent chance that the melting seen over the entire forty-six-year period for which they had data was caused by natural forces alone. When they ran computer models that took into account greenhouse gas emissions, the results very closely matched the observed melting. Their findings are supported by the work of researchers at the Hadley Centre in England. This group developed a computer model for the same purpose and produced similar results.

This image, produced by satellite data and aircraft video, shows water forming on Arctic ice. (Light blue areas show ponds of water on the ice, white areas show sea ice without ponds, and the dark areas are ice-free water.)

CHANGING THE WEATHER

In 2000, Dr. Thomas J. Crowley, a geologist at Texas A&M University, conducted a study into the causes of global warming. He found that in the past, natural causes, such as volcanic activities, were the main source of temperature variations. However, he also found that since 1900, such natural forces have been responsible for only 25 percent of the increase in recorded average temperature.

Dr. Crowley's study was intended to provide insight into the way that natural events were responsible for past shifts in average temperature. For example, there was a warming period during the Middle Ages and a period from the 1600s to the 1800s where temperatures became cooler. Dr. Crowley used a computer model that compared natural forces, such as the amount of light radiated by the sun and volcanic activity, to changes in temperature. He found that the computer model did a good job of matching changes in temperature to variations in these natural phenomena for the period from 1100 to the mid-1800s. However, the relationship between natural forces and temperature broke down when it reached the twentieth century. According to Dr. Crowley's study, the rise in greenhouse gases is the only element that corresponds to the rise in temperatures. Although more work remains to be done to confirm these findings, the study is an important step in understanding the effect of natural and human activities on the climate. Grasping the impact that people have on the climate is the key to addressing potential global warming problems.

The United States National Academy of Sciences issued a 2001 report called "Climate Change Science: An Analysis of Some Key Questions." This report states, "The cooling trend in the stratosphere, evident in radiosonde [a device carried into the air, usually by a balloon, that collects information about weather] data since the 1960s and confirmed by satellite observations starting in 1979, is so pronounced as to be difficult to explain on the basis of natural variability alone." The report goes on to say, "This trend is believed to be partially a result of stratospheric ozone depletion and partially a result of the buildup of greenhouse gases, which warm the atmosphere at low levels but cool it at high levels."

Although these reports are important, scientists are still not certain how much of global warming and changes in climate are the result of human activity and how much is the result of nature forces that vary over time. Still, the sheer volume of data from researchers and computer models suggests that humans are having an impact on the climate. And the nitrogen cycle is one area where such human activity is evident.

PEOPLE AND THE NITROGEN CYCLE

As the number of people on Earth has grown, food production and energy usage have also increased. This has resulted in more nitrogen entering the nitrogen cycle from sources such as the fertilizer used for crops and the burning of fossil fuels. The amount of nitrogen fixation has increased substantially since the middle of the twentieth century.

Food, of course, is critical to human survival. The amount of land used to produce food and the amount of fertilizer used has grown along with the global population. Today, more than 37 percent of Earth's land is used to grow food or as pasture for food animals.

In the past, the most common form of fertilizer was manure. Farm animals ate plants and then produced manure. The manure, in turn, was used to fertilize new crops. Raising the level of nitrogen tends to make plants grow faster. As a result, since the mid-twentieth century, the use of artificial, nitrogen-based fertilizer has increased steadily. In the 2000s, fertilizer production throughout the world is ten times what it was in the 1940s.

Crops absorb nitrogen from this artificial fertilizer just as they do from manure. And when the plants are consumed or the debris from them decays, bacteria return all the additional nitrogen to the atmosphere. This returned nitrogen is in the form of nitrous oxide

In July 2004, farmers removed thousands of dead fish from ponds in Zhengzhou, China. They believe the ponds were polluted by nearby factories.

and other nitrogen compounds.

Nitrogen is also washed from soil, animal waste, and other solid sources by water that flows through the ground. In some areas, so much nitrate is washed from fertilized soil that it has reached toxic levels in the groundwater (the water beneath Earth's surface from which we get our drinking water). The more fertilizer people use to grow crops, the more nitrogen that is carried into the groundwater and the greater the amount that leaches into rivers. As the rivers flow to the sea, the increased nitrogen levels also affect the coastline. A greater growth of water plants such as algae can upset the balance of nature along coasts, leading to the extinction of some organisms.

3 THE NITROGEN CYCLE AND GLOBAL WARMING

Industrial air pollution from factories contributes to global warming.

The burning of fossil fuels is another major source of nitrogen in the environment. Nitrogen oxides are produced when coal or oil is burned. This results in the production of a biologically usable form of nitrogen, nitrate, which is deposited from the atmosphere onto Earth. Burning of fossil fuels is also a major source of nitrous oxide gas, which has increased in the atmosphere over the last few decades.

The rise in the atmosphere's amount of nitrous oxide indicates that there has not only been an increase in nitrogen fixation but also a rise in denitrification, the returning of nitrogen to the atmosphere. Although it is not as common a greenhouse gas as carbon dioxide, nitrous oxide is, in some ways, a more potential danger. In terms of its effects as a greenhouse gas, one molecule of nitrous oxide is equal to 250 molecules of carbon dioxide. Hence, a smaller amount will do more harm. In addition, nitrous oxide lasts for a remarkably long time in the atmosphere without breaking down. A molecule of nitrous oxide can last for 125 years. Therefore, if more and more nitrous oxide is added to the atmosphere, it will lead to an increase in the greenhouse effect over time.

The addition of more nitrous oxide to the atmosphere can also contribute to changes in sea level, rain patterns, and climate. Moreover, larger amounts of nitrous oxide in the atmosphere can deplete its ozone, a gas that protects Earth from the effects of ultraviolet radiation and makes it possible for life to exist.

Let's take a look at some issues that could arise because of nitrogen cycle changes that increase global warming.

COOLING THE OCEAN CURRENT

Ocean water circulates all over the world, moving warm water around the coasts of the continents until

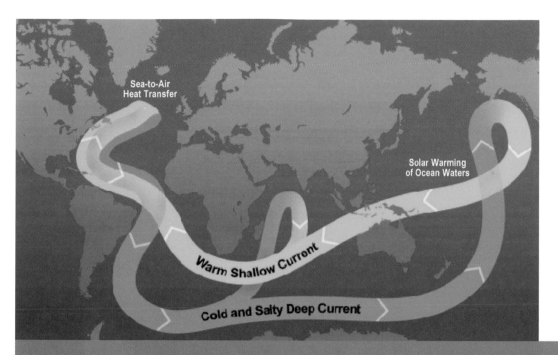

The thermohaline conveyor circulates water around the globe so that warm surface waters from the equator are carried toward the North Pole and South Pole. Deep, cold water warms and rises in the Indian and Pacific oceans and then flows back to the cold, dense waters of the North Atlantic and Antarctic regions, where the cycle begins again.

it gradually cools and sinks. This movement of ocean currents is called the thermohaline conveyor. It is called this because "thermo" means "heat," "haline" refers to salt, and its motion looks like that of a conveyor belt.

The water circulates because the intermixing of warm and cold water creates currents. This process brings warm water from the tropical parts of the Pacific Ocean and the Indian Ocean into the cold North Atlantic

Sea, which washes the northern parts of Europe, the United States, and Canada. The colder water in the north and warmer water in the south drive the ocean current south, and the water continues to circulate. If global warming were to cause a lot of ice melt from Arctic glaciers and the world's mountainous areas, all that cold water entering the thermohaline conveyor could cause it to slow or stop. The warmth brought by the conveyor to countries in the Northern Hemisphere is a major reason why the climate in those areas is relatively mild. Otherwise, being so far north, their climate would be much colder.

If the thermohaline conveyor were to slow, it could lead to harsher winters and a reduction in the moisture found in soil. It could also lead to an increase in wind velocity that would affect areas where much of the world's food is grown, such as the United States and Europe. All of these changes would result in the earth's being able to support fewer people than is now possible.

Evidence from nearly 13,000 years ago, when the thermohaline conveyor stopped functioning, indicates that the climatic effects could last as long as 1,000 years. In a somewhat better scenario, when the conveyor last collapsed 8,200 years ago, the effects lasted only a century. But either way, such a change threatens to make life miserable for a long time.

NITROGEN IN OUR AIR

Besides adding to global warming, nitrogen compounds are a source of air pollution. Nitrogen oxides are gases that contain nitrogen and oxygen in various combinations, such as nitric oxide (NO) and nitrogen dioxide (NO_2). As a group, these gases are referred to as NO_x because of the different number of oxygen atoms they contain. NO_x gases are highly reactive, which means that they combine easily with other elements.

Under the high temperature and pressure in a car engine, oxygen and nitrogen combine to form nitrogen oxide, which enters the atmosphere through car exhaust, above.

These nitrogen gases produce very small particles that travel long distances, hang in the atmosphere, and are a common part of air pollution. Because they are so tiny, they are easily breathed in and stick deep in people's lungs. They also increase the likelihood of respiratory, or lung-based, diseases.

Controlling the production of these nitrogen-based particles is one important goal of pollution control.

About a third of nitrogen oxides comes from industrial pollution, and a large amount comes from car and truck exhaust.

NITROGEN IN OUR WATER

Nitrogen oxides also contribute to water pollution. The nitrogen compound nitrate (NO_3) is washed from soil into the streams, rivers, and other sources of water, and can be carried to reservoirs or other drinking water supplies. This can be dangerous because if people drink polluted water, a chemical reaction can occur in their intestines that changes the nitrate into a different form of nitrogen, called nitrite (NO_2).

The red blood cells that travel through our blood vessels contain a compound called hemoglobin. Hemoglobin attaches to oxygen atoms in the air that we breathe into our lungs. As the red blood cells travel through our body, they deliver this oxygen to all our cells. The cells use this oxygen as fuel to "burn" the nutrients we eat to produce energy for the processes that keep us alive.

Nitrite, however, can change hemoglobin so that it can't carry oxygen. Scientists don't know why, but babies under three months old are especially likely to experience this problem. Because of this danger, the U.S. Environmental Protection Agency (EPA) recommends that drinking water contain nitrate levels below 10 milligrams per liter. In 1998, the EPA estimated that as many as 66,000 babies were drinking water with greater than the recommended level of nitrate. One of the largest sources of water's nitrate pollution is the nitrate washed from farmland where nitrogen-based fertilizers are used. Other sources are industrial waste and sewage treatment plants.

NITROGEN OXIDES AND ACID RAIN

Nitrogen oxides in the air combine with other elements and form acidic compounds, such as sulfur dioxide and nitric acid. When these compounds fall to the earth in rain or snow, it is called acid rain. The compounds can also drift in the air as dustlike particles. Together, the drifting particles and acid rain are called acid deposition. The particles make lakes and streams

Polluted water, above, is being processed at a sewage treatment plant. Pollution in ocean water can kill grasses along the shore and deprive the ocean of oxygen.

Nitrogen and Food Production

The heavy use of nitrogen-based fertilizer to grow food is a contributing factor to global warming. Ironically, global warming could ultimately affect our ability to grow food. Researchers have used computers to predict what the world might be like by 2030 as a result of global warming.

It is likely that current temperatures and rainfall patterns will change. This, in turn, will alter the areas suitable for growing food. The United States has so much farmable land that it is unlikely that its citizens would starve. However, the midwestern United States currently exports food to about 100 countries. By 2030, global warming–related climate changes could cut food production in the American Midwest to a third of what it is today. This would in turn reduce the amount of food being exported to other countries, and people in those countries would suffer.

According to the computer model, although Canada would see an increase in farmable land, it would not be a sufficient amount to make up for the

One of the effects of global warming is that as temperatures increase and lakes dry up, dry areas become even hotter and drier. Africa (*above*) is especially vulnerable to effects of global warming, such as reduced food production, because so many people live at the poverty level.

loss from the United States. Among European countries, Greece and Italy would likely experience the worst effects, while Scandinavian countries would most likely see an increase in their harvests, and Russia would see some Siberian land become more farmable. One of the major problems stemming from global warming is that areas that are already dry and where people are struggling to produce enough food, such as those in Africa, are likely to become even drier.

more acidic and can damage plants that absorb that water. For example, throughout the Appalachian Mountains, deposited acidic compounds appear to be stunting the growth of high-elevation spruce trees. Animals that live in lakes and streams are also affected. In some areas, species such as brook trout have been completely killed off by contaminated bodies of water. Acid deposition can also damage objects such as buildings and cars. Acid rain is an international problem because the acidic particles in the air can be carried by winds to distant locations. Thus, nitrogen pollution can start in one country and affect another country far away.

NITROGEN AND THE ECOSYSTEM

Nitrogen that is washed from fertilized farmland and that falls from the sky as acid deposition is carried into undeveloped land. There it can upset the balance of the ecosystem—the natural system of plants and animals that live in a specific location. The nitrogen acts as fertilizer for the plants that grow in such areas and causes species that use nitrogen most efficiently to grow more rapidly. These species can kill off others, thus reducing the number of different species—or the biodiversity—in an area.

WILTING THE RAIN FOREST

Global warming could also cause trees to die off in both rain forests and temperate areas. Drought and changes in temperature that make the environment unsuitable for growth could cause trees to rapidly disappear. Trees remove carbon dioxide from the air, so if there were fewer trees, more carbon-based gases would remain in the air, making the global warming even worse. This would set up a vicious cycle as increased global warming would kill still more of the trees necessary to contain it. The loss of plant species would likely also lead to a loss of the animals that live on or in them, thus further reducing biodiversity.

4 UNDERSTANDING WHAT'S REALLY HAPPENING

These red pines are located in the Harvard Forest, a long-term ecological research site in Petersham, Massachusetts.

Before any problem can be fixed, it must first be understood. Therefore, scientists study the effects of increased nitrogen on the environment in a number of different ways.

WHAT IF . . . ?

One scientific approach is aimed at developing computer models of exactly how disruptions in the nitrogen cycle are occurring and how they are contributing to environmental changes such as global warming. Models that show what effects these changes are likely to have on the climate also need to be developed.

To do this, scientists program different scenarios into computers. Scenarios are make-believe situations based on a specific set of events. Based on those events, the computer then provides predictions of what would happen.

This type of approach allows scientists to use current activities to predict what will happen in the future. More important, however, it allows researchers to see what will most likely happen if certain behaviors are changed. Will these changes make things better or worse? Is one change better than another? This type of approach is called a "What if?" scenario. The computer scenarios allow us to gain answers to questions without having to use trial and error and then wait decades for answers.

REAL-LIFE TESTING

Some scientists study the effects of elements such as nitrogen on the environment by using small test plots of plants to perform controlled experiments. For example, Susan M. Rainey, Knute J. Nadelhoffer, Whendee I. Silver, and Martha R. Downs published a study called "Effects of Chronic Nitrogen Additions on Understory Species in a Red Pine Plantation," in a 1999 issue of *Ecological*

Since forests are vulnerable to the effects of increased nitrogen levels, research programs attempt to evaluate how they are affected by environmental changes related to human activity. The healthy stand of trees shown here (the control group for a 1999 study) is located in an area of the Harvard Forest where the nitrogen level is neither too high nor too low.

Applications. This study used two stands of red pine in the Harvard Forest in Massachusetts. For seven years, one stand was fertilized with nitrogen. The other was used as a control (an untreated sample to which the experimental sample can be compared). The goal of this study was to see exactly how adding significant amounts of nitrogen to an area changes the kinds of plants that grow there. The researchers found that there were significant negative changes in the types and

These trees are from the 1999 Harvard Forest study. The trees above show the effects of low nitrogen levels and are less healthy than the control group on the previous page. The trees below, which are significantly less healthy than either of the other groups, show the effects of too much nitrogen.

amounts of plants that grow near the ground in the nitrogen-enhanced forest. For instance, a type of fern that was common in the control plot declined in the plot that received the extra nitrogen.

This sample from a tree's roots will be used to measure factors such as how much nitrogen the tree absorbs from the soil.

When scientists simply study the changes in a given area over a period of time, the changes they observe could be the result of any number of factors. The advantage of the experimental approach is that it allows researchers to control exactly what changes are introduced into the system. There is therefore a close link between the changes introduced and the results observed.

SAMPLING THE ENVIRONMENT

Another way that scientists study the effects of increased nitrogen in ecosystems is by taking samples of plants,

Data Mining

Advances in computer technology have allowed many government agencies, such as the EPA, and universities to compile enormous databases of information on a wide range of environmental topics, including pollution and global warming. Such databases can analyze vast quantities of measurements taken over long periods of time. New techniques for analyzing data are continually being developed. One such technique is data mining. In data mining, a computer is programmed to compare extremely large amounts of data and provide information on any important relationships that appear to exist among them. For example, by analyzing a large database obtained by satellite observation, scientists at NASA's Ames Research Center in Moffett Field, California, are using data mining to reveal information about the relationships between human activities, increases in the amount of carbon dioxide in the atmosphere, and climate changes. This type of analysis can also be applied to other pollutants such as nitrogen. Such advanced computing techniques can reveal valuable information in masses of data too large to be analyzed by any other means.

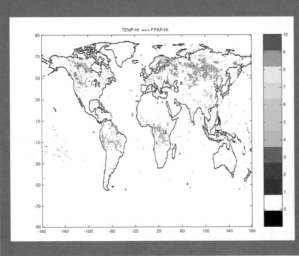

Researchers used data mining to make this map. It shows the geographic areas that are suffering the greatest effects from global warming. Pictured in dark blue, they are the high-latitude northern and Arctic areas of Europe and Asia.

soil, and other materials in a specific area over a long period of time. They note increases in elements such as nitrogen in the soil and the growth of plants in the area. Their observations tell us what changes in ecosystems occur over time as a result of variations in the environment, such as increased nitrogen.

For example, in one study, titled "Major-Element Cycling in a High-Elevation Adirondack Forest: Patterns and Changes 1986–1996," published in *Ecological Applications*, Andrew J. Friedland and Eric K. Miller studied changes that have taken place in a forest in the northeastern United States. The area is highly industrial-ized, and as a result, large amounts of various elements, including nitrogen, have been deposited in its environ-ment. They documented changes such as a decrease in some types of trees like the red spruce, for example, and an increase in others. Long-term studies such as these can reveal valuable information on how plants process elements, how much they absorb, and at what point they have absorbed as much as they can and the excess remains in the environment as a pollutant.

One difficulty with such studies is that the measure-ments must be taken over years or even decades. This means that by the time the effects of environmental changes become evident, it is too late to alter the factors that caused them. However, such data is useful for predicting how other areas might be affected. The results of studying the effects of increased nitrogen availability on the industrialized northeastern United States' forests might provide valuable information for other areas. For example, it can show what might happen in parts of the world just beginning to undergo industrialization or the implementation of modern farming techniques, including the use of nitrogen-based fertilizers.

LEARNING FROM THE PAST

In many scientific areas, including ecology, detailed measurements have been recorded only during the past couple of centuries. As data is compiled over years, it makes it possible to go back to previous records and analyze the changes that have occurred in the past without waiting years to obtain new data. This retrospective analysis can allow scientists to understand how changes in the environment affect it over time. This, in turn, allows potential problems to be addressed before they occur.

5 | THE FUTURE IS IN OUR HANDS

This Oued Tensift riverbed in Marrakech, Morocco, shows the effects of a three-year drought.

What might the future hold? In 2003, Peter Schwartz and Doug Randall wrote a report for the U.S. Pentagon titled "An Abrupt Climate Change Scenario and Its Implications for United States National Security." In it, they summarized some of the changes that could occur if global warming brought about a sudden change in climate. These changes include:

- A drop in temperature of up to 5° to 6°F (about 3°C) across North America, northern Europe, and Asia.

- A rise in temperature of 4°F (about 2°C) in Australia, southern Africa, and South America.
- A decade-long drought in important farming and water-producing areas for major urban areas of Europe and eastern North America.
- An increase in the severity of storms in Western Europe and the north Pacific.

According to the authors of the report, such changes could lead to:

- A shortage of food from a reduction in usable farmland.
- A shortage of water because of drought and changes in rain patterns.
- A disruption in power and energy supplies because of storms.

FIXING THE NITROGEN PROBLEM

The desire to fix the nitrogen problem and knowing what to do are only part of the battle. Far more important is actually *doing* what is necessary to fix it, or at least keep it from getting worse in the future. People can have a positive effect on the environment instead of a negative one, but they must be willing to take action, even if it is inconvenient or costly. The following approaches can help reduce how much humans are affecting the nitrogen cycle and control some of the damage. Some of these things require governmental cooperation. Others are things we can do ourselves.

Establish International Goals

Goals need to be set up worldwide for dealing with the impact humans have on the nitrogen cycle. For example, organizations such as the United Nations can work on creating international goals for addressing the problem.

One objective that could be established is the maintenance of a certain level of nitrogen in specific ecosystems. Achieving such a goal would require a reduction in sources of nitrogen, such as nitrogen-based fertilizer, and the use of technology to remove excess nitrogen from areas that have too much. It would also require either increasing the use of alternative energy sources to reduce the burning of fossil fuels or improving methods of trapping by-products such as nitrogen compounds.

It isn't easy to achieve international agreement to work on reducing the elements that contribute to global warming. In 1997, the United Nations produced the Kyoto Protocol on Climate Change, a treaty that took measures to reduce the output of greenhouse gases. One hundred and fifty countries signed the agreement.

Densely populated Indian cities like Calcutta *(above)* face major pollution problems. Air pollution contains sulfur dioxide, nitrogen oxides, and other pollutant particles. Organizations such as CLEAN-India aim to raise people's awareness about the effects of pollution and steps they can take to help reduce pollution.

However, many industrialized nations, including the United States, which is the largest producer of greenhouse gases, refused to sign because they felt that complying with the limits imposed might hurt it economically. In addition, India and China, which originally signed the agreement, were exempt from meeting the mandatory limits on emissions because they were considered "developing nations." Now these countries are two of the world's fastest-growing economies, and they have indicated that they, too, will not agree to the mandatory limits on emissions.

In 2005, the United States implemented a separate agreement with India, Japan, Australia, China, and South Korea. These countries agreed to work to develop more energy-efficient methods that would reduce greenhouse gases. Unlike the Kyoto Protocol, this agreement didn't include specific target levels to which emissions should be reduced, and it is therefore a weaker agreement.

Control Nitrogen in Fertilizer and Food Production

Since the largest amount of nitrogen enters the ecosystem from fertilizer used to grow crops, it would be a logical first step to reduce the amount of fertilizer used in this process. This reduction would not be without its problems, however. In combination with better seed and improved irrigation, the use of nitrogen fertilizer has been the key to high-yield agriculture. This advance has made it possible to feed a much greater number of people throughout the world, and it has been largely responsible for reducing starvation in many poor countries. Thus, there is a tradeoff between protecting the environment and ensuring the survival of the people who live in it.

A first step in controlling nitrogen from food pro-duction is to document at exactly what points it enters

Around the world, organizations like the International Fertilizer Industry Association have been created to find more environment-friendly manufacturing methods.

the ecosystem. Nitrogen enters the ecosystem from fertilizer absorbed by plants, manure from animals that eat the plants, and food waste from people who eat the plants and animals.

If waste products from animals, people, and industry were used as fertilizer and to produce energy instead of artificial fertilizer and fossil fuels, much of the problem with nitrogen pollution would be solved. It is difficult to achieve such a goal, though. For one thing, it would be costly to process waste products and transport them to farmland where they are needed.

Scientific research has shown, however, that the amount of fertilizer necessary to grow food can be controlled through an approach called precision agriculture. In this method of farming, each field's soil is analyzed to determine exactly which nutrients are needed and in what quantities. Fertilizer is also applied selectively at times in plants' growing cycle when it will be most beneficial, so less is used throughout the entire growing season. Fertilizing in this way can help reduce the amount of excess nitrogen entering the ecosystem.

Harvested plant waste is usually burned in the fields and the chemicals, such as nitrogen, contained in the waste are eventually plowed back into the soil. Scientific research is leading to a better understanding of how much residue from crops should be returned to the soil. Too much can increase the amount of nitrogen in the soil. Research is also being conducted into using plant waste as fuel or to produce energy such as electricity. It is also possible to grow crops specifically for the purpose of producing energy. In the past, alternative fuels have been too costly to win acceptance by individuals and industry. However, as the price of oil and gas continues to rise, such alternative fuels may become desirable not only for ecological reasons but also for economic ones.

REGULATING INDUSTRIAL POLLUTION

The federal government sets limits on how much NO_x can be discharged into the atmosphere by facilities that burn fossil fuels. The amount allowed has decreased in recent years. For example, in 1999, the allowable level in the United States was 219,000 tons (about 200,000 metric tons). In 2003, it was 141,000 tons (about 128,000 metric tons), 35 percent less. In addition, many states, through regulations issued by their individual environmental protection departments, regulate how much NO_x is allowed. Continued emission reductions will also help reduce the amount of nitrogen oxides discharged into the atmosphere.

Emissions Trading

One method used for controlling excess nitrogen produced by industry and large farming corporations is to decide how much total nitrogen to allow into the ecosystem. A cap, or limit, can then be placed on the amount of nitrogen each facility is allowed to produce. Some companies may need to generate more than the allowable limit because the nitrogen they produce is a by-product of their manufacturing processes, and a limit would restrict their quantity of manufactured materials or goods. On the other hand, some companies, in the course of their businesses, may not create any nitrogen.

Companies that do not generate nitrogen can sell their nitrogen-production rights to companies that need to produce more. The right to produce X amount of nitrogen can be sold on the open market similarly to how shares of stock are sold on the stock exchange. In this way, companies can produce varying amounts of nitrogen, but the total amount being returned to the ecosystem remains at the target level. This process is called "cap and trade," and it is a practice that is used today in the United States and the European Union.

CLEANER INDUSTRY

There are a number of things that industry can do to help reduce the amount of nitrous oxide and other greenhouse gases released into the atmosphere. Approaches range from using improved filtering systems to producing more energy from alternative "clean" sources. For example, wind or water power could be used to turn wheel-like devices called turbines, which are connected to generators that convert the energy into electricity.

A type of solar cell that captures heat from the sun and converts it to electrical energy is called a photo-voltaic cell ("photo" means "light," and "voltaic" refers to the ability to produce electrical voltage). Panels composed of such photovoltaic cells are placed in groups and used to create electricity. This method is used in rural areas of many developing countries. For example, in the last few decades, China has brought electricity to areas that did not previously have any. It has used alternative energy sources such as photovoltaic panels and wind power, especially in areas where there was no existing power grid.

A farmer in Lanzhou, China, heats water using a solar panel. China has the world's largest solar energy market. A large portion of the country lacked conventional power grids throughout much of the twentieth century. As China modernized, solar panels have been used both to gather the sun's heat and to generate photovoltaic electric power.

The difficulty in using approaches such as these in industrialized nations is that they are more expensive than most fossil fuels. Many companies would also face the cost of converting their facilities to utilize a different power source. Thus, the use of alternative energy would likely lead to an increase in the cost of products and power to consumers. This, in turn, could make industrialized nations less competitive when selling their goods in markets where developing nations have produced similar products more cheaply. For this reason, industry has often been reluctant to adopt many of these cleaner energy alternatives. However, as the cost of fossil fuels such as oil continues to rise, alternative energy may become more popular.

Better Pollution Control

Better pollution control equipment can also help reduce the amount of nitrogen gases released into the atmosphere. As with any type of fire, burning coal or oil requires oxygen. Oxygen combines with nitrogen to form nitrogen oxide. Therefore, controlling the amount of oxygen used in the burning process is one means of controlling the amount of nitrogen oxide produced. Power plants and industrial companies that burn fossil fuels face the issue of how to reduce nitrogen oxides in a cost-effective way. For example, Praxair, Inc., of Danbury, Connecticut, has developed an oxygen-enhanced low NO_x system. It controls how much oxygen is introduced into the burning process and therefore reduces by-products. It can also be used along with other methods of pollution reduction to further limit emissions.

YOUR ROLE AS A CONSUMER AND CITIZEN

We have an impact on how much nitrogen is placed in the environment both by how and what we choose to

Painting Away NO$_x$

Since vehicle exhaust is a major source of nitrogen oxides, one of the best ways to reduce their output is to find some way to treat the exhaust. Millennium Chemicals, a company in the United Kingdom, is working on just that. They are developing Ecopaint, which can be used on buildings and building materials. This paint contains tiny particles of the chemicals calcium carbonate (a compound that makes acids such as nitric acid chemically inactive) and titanium dioxide. When molecules of nitrogen-based gases in vehicle exhaust come in contact with the paint, they stick to the titanium dioxide particles. The particles absorb ultraviolet radiation from sunlight. They then use the energy from this radiation to convert the potential polluting nitrogen molecules into nitric acid. This nitric acid is either naturally washed away by rain water or it can be treated with more calcium carbonate to eliminate, or neutralize, its acidic properties, making it harmless. Ecopaint is being tested through the European Photocatalytic Innovative Coverings Applications for Depollution Assessment project.

Since it still produces some nitrogen compounds and carbon dioxide, another greenhouse gas, this paint doesn't address the entire problem of vehicle exhaust pollution. It could, however, significantly reduce the nitrogen oxide gases in the atmosphere, which not only contribute to the greenhouse effect but can also cause serious lung problems. It is an example of how new approaches to the problem of excess nitrogen production can be developed to also solve the problems of pollution.

eat and by how we choose to manipulate the food we grow. For instance, genetic engineering is increasingly used to grow food with desirable characteristics. There is some controversy and question about the impact that a genetically engineered species of plant will have on the environment. For instance, if a type of corn is genetically engineered to fix nitrogen, a process that now occurs naturally only in legumes such as beans, this may lead to larger corn. However, producing vast amounts of such corn may result in further overloading the environment with reactive nitrogen compounds.

The choices we make in other areas of our lives also have an impact on the production of nitrogen oxides and other greenhouse gases. Lawn fertilizer, for example, is a source of nitrogen pollution. Choosing to use less, or none at all, is an action that people who have lawns can take. Choosing to drive a fuel-efficient car or a hybrid car that runs on electricity as well as gas can help. So can choosing energy-saving appliances and electronic devices, such as computer monitors with the Energy Star logo. The more we work together to reduce the amount of consumed fossil fuels and the more we support improved technology that reduces the output of nitrogen-based and other greenhouse gases, the better life will be for all of us in the twenty-first century.

GLOSSARY

acid deposition The falling of acid-containing compounds (dust and acid rain) from the atmosphere.

acid rain Rain that contains acids formed in the atmosphere by water combining with chemicals from gases such as nitrogen oxides.

atom The most basic particle from which matter is made.

biodiversity The range of living things present in a given ecological community or system.

compound A chemical substance formed from two or more elements.

computer model A computer program that uses a built-in set of rules to predict the results of a process or to simulate how a given set of conditions will change over time.

control group A group of test subjects in an experiment who are left untreated or unexposed to a procedure and then compared with treated subjects.

denitrification The breaking down of compounds so that nitrogen is released into the atmosphere.

deoxyribonucleic acid (DNA) The compound that is the major component in the chromosomes of a cell and carries genetic information.

dinitrogen The most abundant form of nitrogen in the atmosphere, made up of two molecules of nitrogen bound together.

ecology The study of the relationship between living organisms and their environment.

ecosystem The living things in a particular environment.

genetic engineering A branch of biology dealing with the alteration and recombination of genetic material by technical means in order to produce new species with enhanced desired characteristics.

greenhouse gases Gases such as carbon dioxide and nitrous oxide that contribute to global warming.

groundwater Water beneath Earth's surface from which drinking water is obtained.

molecule A combination of atoms that forms the smallest amount of a compound with identifiable properties.

nitrogen fixation The process of converting dinitrogen into a form that can be used by living organisms.

ozone A highly reactive gas that contributes to global warming.

precision agriculture A form of farming in which the exact amount of nutrients required to grow plants is identified and used.

radiosonde An instrument carried into the air by a balloon and used to measure and transmit meteorological data by radio.

temperate Having a moderate climate without extremes in temperature.

wavelength The distance between repeating units of a wave pattern.

FOR MORE INFORMATION

Environment Canada
70 Crémazie Street
Gatineau, Quebec K1A 0H3
Canada
(800) 668-6767 or (819) 997-2800
Web site: http://www.ec.gc.ca

Environmental Defense Fund
257 Park Avenue South
New York, NY 10010
(212) 505-2100
Web site: http://www.environmentaldefense.org

Sierra Club
85 Second Street, 2nd Floor
San Francisco, CA 94105
(415) 977-5500
Web site: http://www.sierraclub.org

U.S. Environmental Protection Agency (EPA)
Ariel Rios Building
1200 Pennsylvania Avenue, N.W.
Washington, DC 20460
(202) 272-0167
Web site: http://www.epa.gov

WEB SITES

Due to the changing nature of Internet links, the Rosen Publishing Group, Inc., has developed an online list of Web sites related to the subject of this book. This site is updated regularly. Please use this link to access the list:

http://www.rosenlinks.com/eet/clch

FOR FURTHER READING

EarthWorks Group, The. *50 Simple Things Kids Can Do to Save the Earth*. New York, NY: Andrews & McMeel, 1990.

Fridell, Ron. *Global Warming*. New York, NY: Scholastic, 2002.

Silverstein, Alvin, Virginia B. Silverstein, and Laura Silverstein Nunn. *Global Warming*. Minneapolis, MN: Lerner, 2003.

Suzuki, David T., and Kathy Vanderlinden. *You Are the Earth: Know the Planet So You Can Make It Better*. Vancouver, B.C.: Sterling, 2002.

Taylor, Barbara, and Scoular Anderson. *How to Save the Planet*. Danbury, CT: Franklin Watts, 2001.

BIBLIOGRAPHY

Advanced Coatings & Surface Technology Staff. "Novel Paint for NO$_x$ Reduction." *Advanced Coatings & Surface Technology*, Volume 17: Issue 3, March 2004.

Ehereman, Gayle. "Mining What Others Miss." *Mechanical Engineering*, February 2005. Retrieved August 15, 2005 (www.memagazine.org/backissues/ feb05/features/miningwh.html).

Friedland, Andrew J., and Kevin Miller. "Major-Element Cycling in a High-Elevation Adirondack Forest: Patterns and Changes 1986–1996." *Ecological Applications*, Volume 9: Issue 3, August 1999.

Jordan, Thomas E., and Donald E. Weller. "Human Contributions to Terrestrial Nitrogen Flux: Assessing the Sources and Fats of Anthropogenic Fixed Nitrogen." *BioScience*, Volume 46: Issue 9, October 1996.

Mlot, Christine. "Tallying Nitrogen's Increasing Impact." *Science News*, Volume 151: Issue 7, February 15, 1997.

National Academy of Sciences. *Climate Change Science: An Analysis of Some Key Questions*. Washington, D.C.: National Academy Press, 2001. Retrieved July 23, 2005 (http://books.nap.edu/html/ climatechange/5.html).

Oil Daily Staff. "New NO$_x$ Reduction Rules Impact Northeast Fuel Markets." *Oil Daily*, Volume 53: Issue 90, May 12, 2003.

Rainey, Susan M., Knute J. Nadelhoffer, Whendee L. Silver, and Martha R. Downs. "Effects of Chronic Nitrogen Additions on Understory Species in a Red Pine Plantation." *Ecological Applications*, Volume 9: Issue 3, August 1999.

Revkin, Andrew C. "Scientists Now Acknowledge Role of Humans in Climate Change." *New York Times*, October 26, 2000.

Schwartz, Peter, and Doug Randall. "An Abrupt Climate Change Scenario and Its Implications for United States National Security." U.S. Pentagon, October 2003. Retrieved June 12, 2005 (www.ems.org/climate/pentagon_climatechange.pdf).

Skirble, Rosanne. "American Geophysical Union Takes Strong Stand on Global Warming." Retrieved July 23, 2005 (http://www1.voanews.com/article.cfm?objectID=BAB97DrC-169C-42).

Socolow, Robert H. "Nitrogen Management and the Future of Food: Lessons from the Management of Energy and Carbon." *NAS Colloquium: Plants and Population: Is There Time?* Beckman Center of the National Academy of Sciences, University of California, Irvine, California, December 5–6, 1998.

Sprent, Janet I. *The Ecology of the Nitrogen Cycle.* New York, NY: Cambridge University Press, 1987.

Suplee, Curt. "Study: Arctic Sea Ice is Rapidly Dwindling; Global Warming Called Likely Cause." *Washington Post*, December 3, 1999.

Tickle, Amy. *Ecology and the Environment: A Look at Ecosystems of the World.* Ann Arbor, MI: University of Michigan Press, 1998.

United Nations Environment Programme/GRID-Arendal. "Potential Impacts of Climate Change: Great Ocean Conveyor Belt." Retrieved June 12, 2005 (http://www.grida.no/climate/vital/32.htm).

U.S. Environmental Protection Agency. "Air Trends: Continued Progress Through 2004." Retrieved August 5, 2005 (http://www.epa.gov/airtrends/2005/econ-emissions.html).

INDEX

ABOUT THE AUTHOR

Jeri Freedman has a B.A. from Harvard University. She has worked in the biomedical and high-technology fields for the past fifteen years, including at companies that provide environmental testing services and manufacture environmental testing equipment. She lives in Boston.

PHOTO CREDITS

Cover © AP/Wide World Photos; p. 1 © Chinch Gryniewicz/Ecoscene/Corbis; p. 4 © James L. Amos/Corbis; p. 8 © Reza/Webistan/Corbis; p. 11 © Adam Hart-Davis/Photo Researchers, Inc.; p. 12 © Washington State Dept. of Ecology; p. 13 © SSPL/The Image Works; p. 14 © Craig Tuttle/Corbis; p. 16 © Ed Kashi/Corbis; pp. 17, 43 © AP/Wide World Photos; p. 18 © Tahara Anderson; pp. 20, 40 © NASA; p. 24 © STR/AFP/Getty Images, Inc.; p. 25 © Simon Fraser/Photo Researchers, Inc.; p. 27 © National Assessment Synthesis Team, Climate Change Impacts on the United States: The Potential Consequences of Variability and Change (Washington, DC: U.S. Global Change Research Program, 2000); pp. 29, 31 © Getty Images; p. 32 © Anthony Bannister/Photo Researchers, Inc.; pp. 35, 39 courtesy of Alison Magill; pp. 37, 38 courtesy of Christian Arabia; p. 46 © Prat Thierry/Corbis Sygma; p. 48 © Robert Brook/Photo Researchers, Inc.; p. 52 © Frederic J. Brown/AFP/Getty Images, Inc.

Designer: Thomas Forget; Editor: Liz Gavril; Photo Researcher: Hillary Arnold